by Amanda Moore

SCHOOL PUBLISHERS

Printed in China

ISBN 10: 0-15-358391-6
ISBN 13: 978-0-15-358391-9

Ordering Options
ISBN 10: 0-15-358356-8 (Grade K On-Level Collection)
ISBN 13: 978-0-15-3518356-8 (Grade K On-Level Collection)
ISBN 10: 0-15-360544-4 (package of 5)
ISBN 13: 978-0-15-360644-1 (package of 5)

5 6 7 8 9 10 985 15 14 13 12 11 10 09

my eyes

my nose

my ears

my mouth

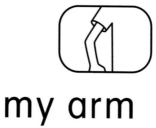

my arm

my hands

my feet